Methane Production Guide

How to make biogas. Three simple anaerobic digesters

Originally published by OneToRemember 2012

This version published by EnergyBook - 2020

ISBN-13: 978-1468161502

ISBN-10: 1468161504

Richard W Jemmett

EnergyBook.co.uk

For more information on Methane Production visit

EnergyBook.co.uk/methane

V3

Contents

Methane – Biogas Production Guide

Preface

This book describes how you can produce methane from organic materials and waste. Methane produced by the anaerobic digestion process is often known as biogas and is quite similar to natural gas that is extracted from the wellhead and piped to our homes. However, natural gas contains a variety of hydrocarbons other than methane, such as ethane, propane, and butane giving it higher heat content (calorific value).

Once you have understood the basic principles of methane production in this book, you will be able to create more elaborate equipment from the designs described that will make useful amounts of gas.

Anaerobic digestion is one of the most common chemical processes in nature. Anaerobic means the decay or breakdown in the absence of air or more specifically oxygen. From the designs in this manual you will be able to move from a simple table top experiment through a batch unit to a continuous output digester.

Methane gas when mixed with air (oxygen) is highly explosive and every safety precaution must be taken when constructing and using this equipment. Safety should always come first. If after reading this manual you believe that you do not have the right skills to build and operate a methane generator safely please do not continue.

Basic properties of methane

Methane is a colourless, odourless, flammable[i] gas and the main constituent, 85% to 90%, of the piped natural gas[ii] that we use in our homes in the UK, Europe and the USA. Its chemical symbols are CH_4 and it is a hydrocarbon.

Methane produced by the anaerobic digestion process is quite similar to natural gas that is extracted from the wellhead and piped to our homes. However, natural gas contains a variety of hydrocarbons other than methane, such as ethane, propane, and butane. As a result, natural gas will always have a higher calorific value than pure methane. Depending on the digestion process, the methane content of biogas is generally between 55%-80%. The remaining composition is primarily carbon dioxide, with trace quantities (0-15,000 ppm) of corrosive hydrogen sulphide and water.

The average expected energy content of pure methane is 33.4-39.8MJ/m^3 (896-1069BTU/ft$^{3)}$; natural gas has an energy content about 10% higher because of added gas liquids like butane mentioned above. However, the particular characteristics of methane, the simplest of the hydrocarbons, make it an excellent fuel for many uses. With some equipment modifications to account for its lower energy content and other constituent components, biogas can be used in energy consuming applications designed for natural gas.

The gas made using the suggested ideas contained in this book will be made up of methane plus other gasses or 'diluters' and have a typical value of 22.3MJ/m^3 (600BTU/ft^3).

Anecdotal evidence indicates that biogas was used for heating bath water in Assyria during the 10thcentury BC and in Persia during the 16th century. Jan Baptita Van Helmont first determined in the 17th century that flammable gases could evolve from decaying organic matter. Count Alessandro Volta concluded in 1776 that there was a direct correlation between the amount of decaying organic matter and the amount of flammable gas produced. In 1808, Sir Humphry Davy determined that methane was present in the gases produced during the anaerobic digestion of cattle manure.

The first digestion plant was built at a leper colony in Bombay, India in 1859. Anaerobic digestion reached England in 1895 when biogas was recovered from a 'carefully designed' sewage treatment facility and used to fuel street lamps in Exeter. The development of microbiology as a science led to research by Buswell and others in the 1930s to identify anaerobic bacteria and the conditions that promote methane production.

Methane gas also occurs naturally as swamp gas produced from murky stagnant water. Lightning could ignite this gas and has been said to be the origin of Willow the Wisp. Indeed the term swamp gas is often used for methane gas produced by anaerobic digestion. Landfill gas is also made up of a high proportion of methane and is a good example of the commercial use of methane production where the gas is often used to drive a gas engine and electric generators.

Gas made with a digester is also commonly called biogas.

Advantages
- Can make use of organic wastes and a fertilizer is produced at the end of the process.
- Is a clean, easily controlled source of renewable energy.
- Reduces pathogen (disease agent) levels in the waste.
- Equipment can be simple to build and operate.
- Low maintenance requirements.
- Can be efficiently used to run cooking, heating, gas lighting, absorption refrigerators and gas powered engines. Unlike solar PV and wind turbines, biogas is a good form of renewable energy for heating.
- No smell (unless there's a leak, which you'd want to know about and fix immediately anyway!).

Disadvantages
- Most practical when used at the source of the waste. This is because the energy needed to compress the gas for transport, or convert it into electricity is excessive, reducing the overall efficiency of biogas energy production. However in the UK there are large scale commercial plans to inject biogas (biomethane) into the gas distribution system.
- For safety, basic precautions must be adhered to. Biogas is extremely dangerous in generation and use.

Anaerobic digestion

Anaerobic digestion is one of the most common chemical processes in nature. Anaerobic means the decay or breakdown in the absence of air or more specifically oxygen. The process is similar to fermentation as the transformation is brought about by micro-organisms (bacteria) called anaerobes. Like with the production of alcohol (ethanol) digestion takes place in two stages. First, in the medium of digestion certain micro-organisms break down the materials into simple sugars, alcohol, glycerol and peptides. When these components are present in the correct amounts and the conditions are correct, a second group of micro-organisms converts these simpler molecules into methane gas.The micro-organisms are particularly sensitive to environmental conditions including temperature and acidity.

Anaerobic digestion occurs between 0oC (32oF) and 66oC (150oF). However the optimum temperature which promotes activity of the micro-organisms and consequently produce more methane gas is between 30oC (85oF) and 35oC (95oF). In colder climates this is difficult to maintain but worthwhile trying to achieve. Below 60o F little gas is produced.

Acidity is also important with a desired pH of between 7 and 8. With a low acid content (high pH) the fermenting slows down until the bacteria produce enough acid (acidic carbon dioxide) to restore the balance. Acidity can be measured using litmus paper (can be purchased from Amazon).

Methane – Biogas Production Guide

Carbon and nitrogen are the other two components for a digester and are both required for the micro-organisms to live. However, the bacteria consume the carbon at about 30 times faster than the nitrogen. This 30:1 ratio produces the maximum amount of gas. If the ratio is not correct the bacteria will usually compensate creating the right balance within the digester.

As mentioned earlier the gas produced in a digester is not pure methane and is usually 75% methane and 25% carbon dioxide (CO_2) with trace amounts of hydrogen, nitrogen and other gases characteristic of the original materials used in the digester.

The slurry that is left after the digestion process is complete is mainly composed of organic humus, with small amounts of nitrogen and phosphates. This final product of gas production makes an excellent fertiliser and soil conditioner.

It should be noted that the time in starting the digester and producing gas can be as long as four weeks - but sometimes as short as two weeks. This is because the bacteria will first need time to breakdown the slurry into alcohols and sugars, before the second group of bacteria, the gas producing ones, can adjust the carbon/nitrogen mix and the acidity level for reasonable amounts of gas to be produced.

Modest experiment in methane gas production

Having read the first part of this guide many readers may want to build a methane digestion plant now and power up their houses. However, while you are waiting to build a digester large enough to process your household waste and other peoples' waste from down the street into enough methane to heat the house; you may wish to try a simple, low cost experiment. This will help familiarise you with the fuel's production and some of its characteristics.

Here is how to put together one of the simplest and least expensive methane production experiments of all. You will need only a gallon cider jug, some sort of gas holder (a recycled, heavy-duty plastic bag) and, from the chemistry lab, some rubber tubing, a couple of tubing clamps, a two-hole rubber stopper, glass tubing and a glass "Y".

Your first step in constructing a mini-methane-generator will be to make a manometer. This is a U-shaped tube, partly filled with water, that will let you know when your little digester is producing gas, indicate the pressure of that gas and act as a safety valve (since excess pressure will blow the water out of the manometer). Any chemistry student should be able to show you the proper way to heat and form your glass tubing.

How to bend glass tubing

It is useful to know how to bend a piece of glass tubing, especially if you are interested in chemistry and want to set up some apparatus.

There are just two things you need to bend glass tubes with and these are (1) a Bunsen burner and (2) the glass tubing.

This is a burner in which a jet of ordinary illuminating gas is mixed with air, the amount being regulated by a ring which opens and closes the air holes in the burner.

A Bunsen burner makes a very hot flame because the gas in the tube moves faster than in an ordinary burner and the oxygen in the air aids the gas to burn hotter. If you have no gas in your house you can use an alcohol lamp which you can either buy or make for yourself.

To bend a piece of glass tube you should have a fish tail jet set in the end of the Bunsen burner to give a wide flame like an illuminating burner. Hold the tube over the flame of the burner, or alcohol lamp until it is heated red hot all along the place you want to bend it.

Now turn the tube in the flame with your fingers until it is heated evenly all around and becomes soft; take it from the flame and quickly but gently bend it as you wish which you can do very easily. With very little practice you will be able to make a good smooth bend just where you want it.

The four inch manometer dimension shown in the drawing should be considered a maximum for both practical and safety reasons. Filling the tube with water to such a depth will give you eight inches of pressure (eight inches water gauge is about the same pressure as the gas in most UK homes downstream of the meter) and therefore more than sufficient. Gas appliances usually operate on pressures of less than eight inches and there is no reason for you to risk blowing your jug apart with gas compressed beyond this amount.

Once your manometer is completed, you should make a "burner tip" by drawing out a piece of glass tubing in the approved manner (again, any chemistry student should be able to help you if you have never formed glass tubing before). The tip should be quite long as a precaution against the possibility of a back flash. Then attach the stretched-out burner to one arm of your glass "Y" with a short piece of rubber tubing on which a clamp is placed to act as a valve.

The other branch of the "Y" feeds directly to your gas collector through a longer section of rubber tubing (also fitted with a clamp). The experiment that wrote this article made a collector from a polyethylene milk bag taken from a cafeteria-type dispenser. The cardboard cartons that fit inside such dispensers are thrown out after one use and you will find that each box contains a bag-liner. Fully inflated, the bags are somewhat larger than a king-sized pillow. Wash one out, roll it up to expel the air inside and hook it to the "Y".

Now you are ready to place some manure in the jug. The best type appears to be a mixture of droppings and litter from chickens but, if you cannot get that, try something else such as straight horse manure. As mentioned earlier in this guide the very most efficient formula is 30 parts of carbon to one part nitrogen.

Mix the manure with water to form slurry and pour it into the jug. Fill the jug to about four inches below the stopper (there will be some initial foaming and you want to keep it out of the tubing).

Once again the most efficient generation of methane takes place at 32°C (90°F) to 38°C (100°F) and, if your slurry's temperature drops much below 27°C (80°F), the gas production will be slow or nonexistent. You will have to provide a sufficiently warm environment for your jug, then, if you want it to make gas. Bear in mind, though, that methane— carelessly handled—can explode so take suitable precautions in setting up your apparatus. Never place the experiment near a naked fire or burner. Never use a jar or tin to collect the gas in as it will contain air and therefore an explosive mixture will be made. Always use a bag or something similar from which all the air can be expelled.

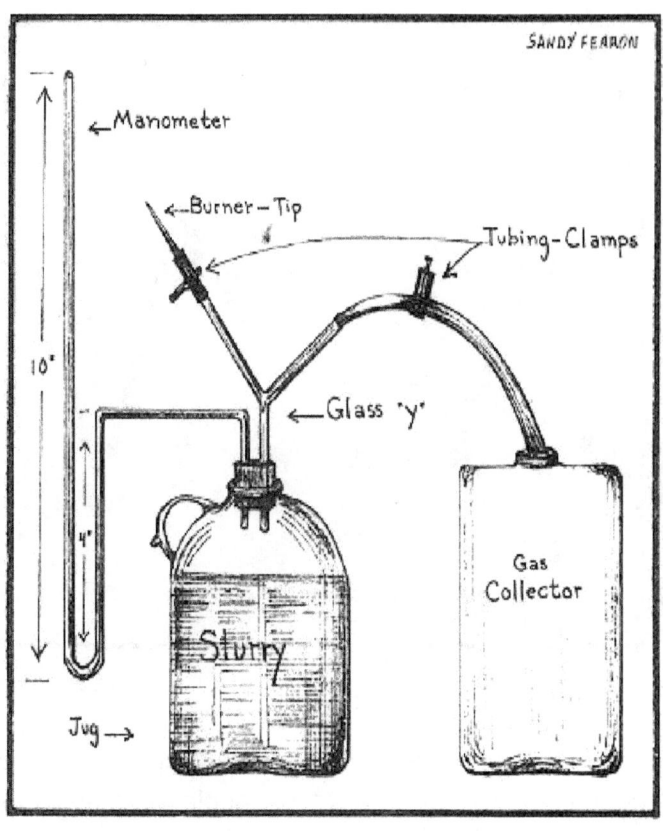

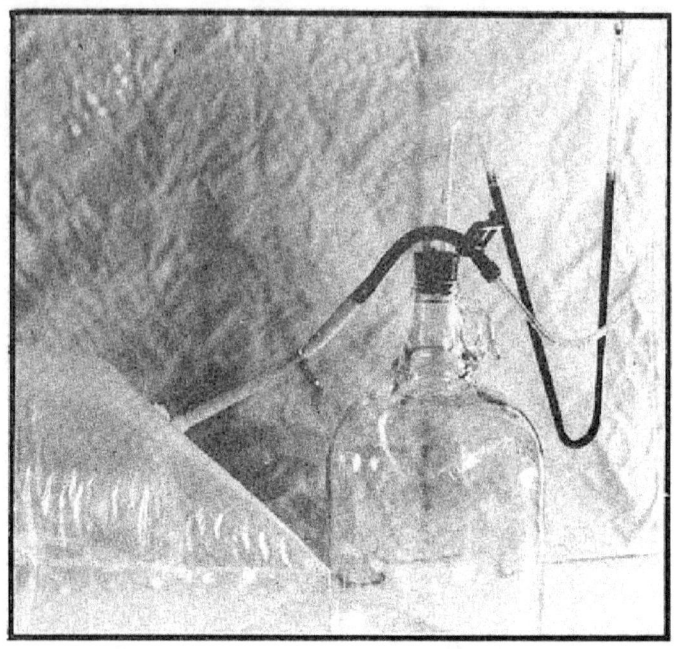

Start your generator working with all its valves (clamps) closed and, after a couple of days, the water being "pushed" up the long arm of the manometer will indicate that some pressure is beginning to build in the jug. This first production is mostly carbon dioxide, which will not burn. (Test the gas by holding an ignited match at the tip of the burner and opening its clamp. The amount of gas in the manometer is sufficient for such a trial, although —as stated—the carbon dioxide will not burn.)

Continue the tests until a match held at the burner tip does ignite the escaping gas. This may take a couple of weeks or more depending upon the acid conditions of the slurry in your jug.

Eventually, incorrect acidity levels will correct themselves and your model generator will begin to produce methane. When you are satisfied that such production is underway, open the clamp to the gas collector and you're in business. Methane production —depending on temperature—should last for from one to three months.

What can you do with the gas? You can burn it off through the burner tip as a graphic demonstration that decomposed organic matter really does produce usable fuel. The quantity is too small for much else. To increase the pressure of the escaping gas (and, thereby, the spectacular nature of the resulting flame), place one or more weights on the collector bag. The manometer, of course, will faithfully indicate the pressure your gas reaches during such a demonstration.

Once the thrill of watching the flame passes, disconnect the collector bag, take it outside and expel the remaining methane. Remember the residue left in the jug is an excellent fertiliser and you can use the liquid and some of the solids to seed your next batch of waste (and thereby hasten its production of gas).

The author also suggests a couple of untried refinements. If you have a fish aquarium heater available, you might try putting your jug in a bucket of water warmed by the element. This would be a significant improvement in maintaining the digesting slurry at optimum working temperature. You can also improve the burning qualities of the resulting methane by bubbling it through a lime water solution to remove carbon dioxide and passing it over ferric oxide (rust) to remove hydrogen sulphide.

Although the above experiment is imprecise and yields only a small quantity of methane, it will familiarise you with the digestion process and, possibly, encourage you to investigate the construction of larger-scale generators that will produce usable quantities of gas.

Simple anaerobic digester

A larger digester can be made very simply. The main component required to make a simple digester is the vessel that will contain the slurry. The vessel must allow a method of filling as well as a way of extracting the gas. The following diagram is a simple digester for demonstration purposes only, and is not meant for construction in its exact state as, although correct in concept it does not allow for the safe storage of the gas produced.

Gas storage

Once the gas has been produced a practical and safe means of storing the gas is essential. As discussed earlier the storage container must not contain air and there must be no way for air to enter the storage vessel. Therefore the vessel must have absolutely no air in it before the gas is introduced. If not an explosive mixture will be produced. It is never adequate to allow a gas air mixture to be formed and rely on there being no source of ignition. Sources of ignition can arise from static electricity, for example, and therefore there is always a potential source of ignition nearby.

This basic digester will produce a modest amount of methane gas. It is a good model to try out to become familiarised with the process of methane production.

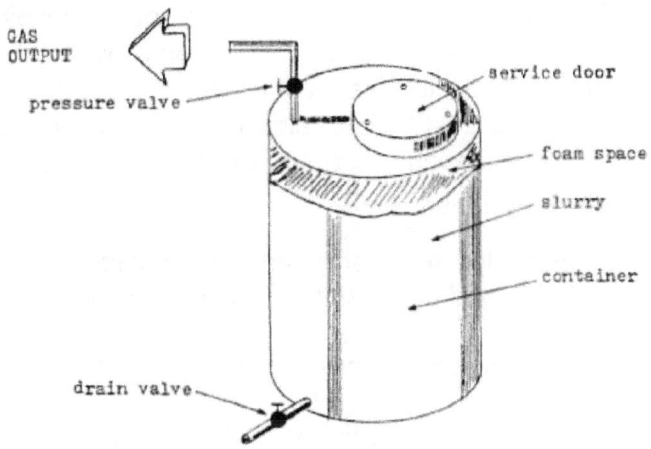

This type of digester is known as a batch feed system, where slurry is introduced into the digester through a service door that is then sealed closed. After a few weeks once the conditions are right, fermentation begins. An airspace at the top of the vessel to allow the first group of bacteria some oxygen to breakdown the slurry into simple molecules and to help prevent foam produced during the digestion process to travel into the pipes. After a couple of months the batch will no longer produce gas. At this point the drain valve is opened and the decomposed matter removed. The vessel may be flushed through but a small amount of slurry should be kept to help start the next batch.

Batch digester construction

This little digester will provide enough free gas to provide heat to cook one meal a day. Modest applications like lighting small rooms with gas lanterns and cooking are ideal for this system. It can also be a low cost build. A simple inner tube from a large tyre such as a tractor will make a perfect container for the gas as it:

- Is a relatively inexpensive and easy to obtain
- May be purged of air relatively easily by rolling tightly
- Automatically creates pressure for feeding the appliance
- Is about the right volume for the size of the digester.

Purging the container or pipe means to remove all air from it. It cannot be emphasised enough that never must air be allowed to mix with the methane.

The container used in this system is a standard 44-gallon oil drum. Try to get one that is relatively clean with no rust whatsoever. Safely remove any residues in the tank with soap and water and then clean water. Oil drums are used to store a very wide range of chemicals and oils so find out what was stored in it before you flush out the contents. If in doubt look for another container that has been used for a safer product.

A stable and secure base can be made out of a few concrete bricks or slabs. The drum when full will be very heavy. The container should be kept off the ground to prevent rusting where possible. The drum should also be high enough to allow draining into a suitable container.

There are normally two vent holes at the top of the drum. It is best to try to use these to fill the vessel. They will need to be closed afterwards and be gas tight. If a larger access hole is required one suggestion is to use an air filter cover from an old car; some had large metal air filters with one or two bolts to secure them and a rubber gasket to provide an air tight seal. After cutting out the right size hole a cross bar could be fitted across the hole with bolts aligned to fit the air filter cover.

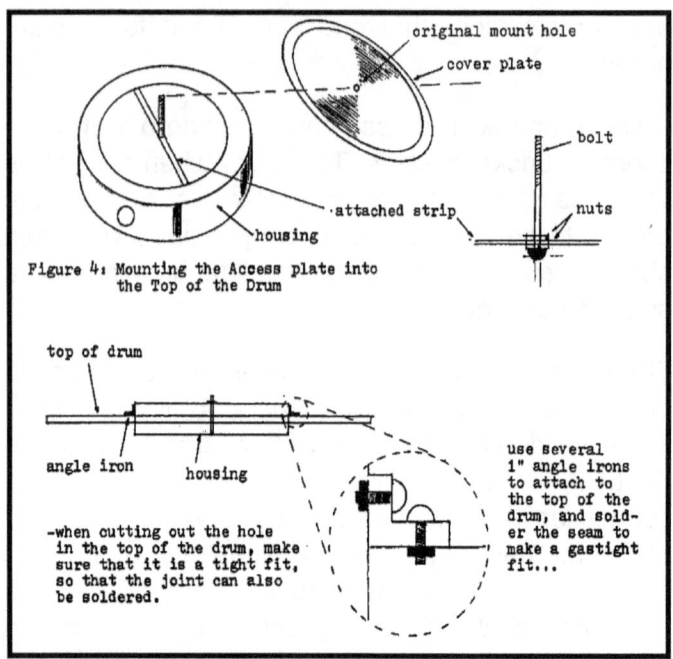

original mount hole

cover plate

bolt

nuts

attached strip

housing

Figure 4: Mounting the Access plate into the Top of the Drum

top of drum

angle iron   housing

-when cutting out the hole in the top of the drum, make sure that it is a tight fit, so that the joint can also be soldered.

use several 1" angle irons to attach to the top of the drum, and solder the seam to make a gastight fit...

Last, drill a snug fitting ¾ " hole into the top of the drum and install the gas outlet pipe. A further hole should be drilled near the bottom of the drum so that a drain can be fitted with a valve. The materials for these pipes can be iron or copper. Plastic pipe could be used for the drain. However sometimes this pipe can go brittle when exposed to sunlight.

For larger outputs two units can be constructed in series. Indeed as the units are simple and cheap to build you may wish to build even more.

Suggested parts list
44 gallon oil drum

Air filter housing for service door, optional

Concrete bricks for base

¾ " copper piping

'T' joint with 1" copper reducer

Valves

Large tyre inner tube

Tyre hose, screw on type Iron work for service door

Copper fittings-compression fittings are not recommended PTFE tape, jointing compound, etc. Solder and propane blow torch.

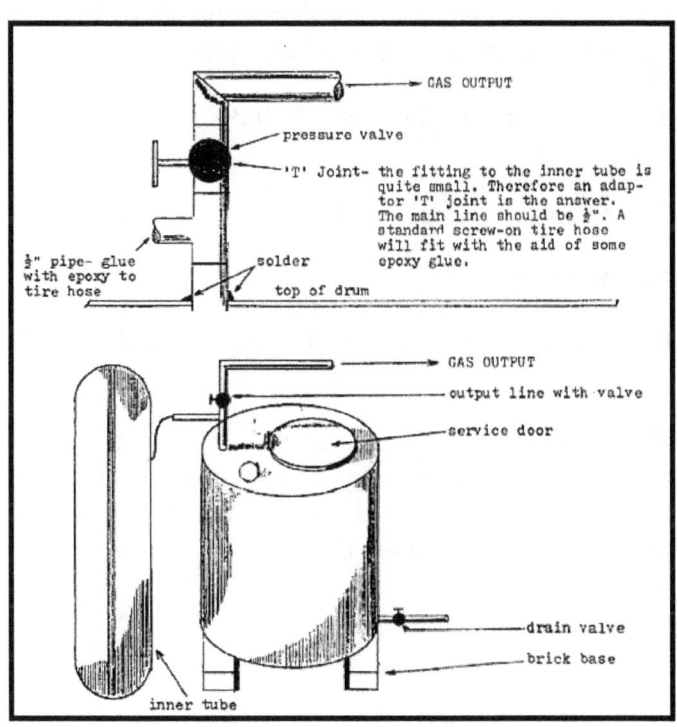

Digester operation

The composition of the slurry will to a large extent determine the success of your digester. To get the 30:1 ratio of carbon and nitrogen animal manure appears to be best. Adding grass cuttings and leaves maybe acceptable but they contain little or no nitrogen. Trial and error may help you find the right mix.

On a farm, manure is readily available but in the city less so. It is then possible to mix leaves and grass clippings with organic waste from the kitchen. This can include fruit and vegetable peelings but not cooked food, meat, paper or cardboard.

Ideally the slurry that works best in the digester comprises:

3 to 4 gallons of liquefied manure

10 gallons of water

Enough grass cuttings and leaves (50:50 ratio) to fill the vessel within 1 foot of the top.

The mixture should be stirred well and should produce gas after about 2 weeks with peak production after about 8 weeks. There will be little production after 12 weeks.

When gas is being produced - try bubbling the output through some water rather than into the storage vessel. Leave it to produce gas for several days until you are certain that all the air has been expelled. *DO NOT LIGHT THE GAS.* The vessel and pipe work may still contain air and therefore you might cause an explosion.

Then take the inner tube and remove the tyre valve. Roll the tube very tightly pushing all the air out of the tube. When this is complete replace the valve and screw the valve onto the 'T' joint. The system should now be free of any air and ready to accept an appliance.

You must also purge all pipes that are added to the system at this point as they will have air in them until the gas passes through. You can let the appliance run for a few minutes before lighting when it is first connected. Do not do this in an enclosed area where the venting gas can build up.

Digester performance

This type of batch feed system does offer some drawbacks as you will probably need the gas each day rather than waiting for two weeks. One solution is to use two digesters and aim for one to reach peak performance whilst the other is fermenting and starting to produce gas. The following diagram shows the performance of a one drum digester and two drums in series.

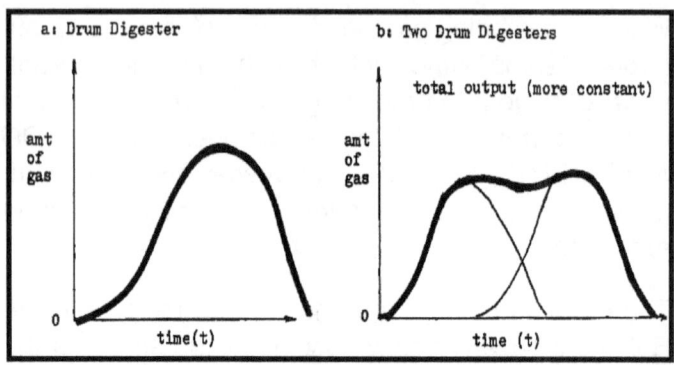

Continuous output digester

A fairly large digester can be built from two 275-gallon boiler oil tanks. One can be used for the digester and one for the holding tank.

A feed chamber can be placed at the end of the digester, with an airtight valve at the top and bottom of its column.

The exhaust tube can be placed up near the top of the tank, but low enough so that the level of the used slurry flows out of the digester (about 8 gallons) will come out. On the other hand a pipe too low will be exhausting slurry that is still digestible. A simple way of setting the correct height is to add exactly 8 gallons of liquid slurry when filling at the point when the slurry just starts to overflow out of the exhaust pipe. Close the valve and add 8 more gallons.

The holding tank should be equipped with a pressure valve measuring up to 50 psig. The pressure of the gas should be monitored closely and any excess gas vented or consumed.

The holding tank cannot of course be collapsed. Therefore, a displacement method must be used to purge it of air. Filling the tank with water to the very top ensuring that there is no air present can do this. Once the feed line from the digester is purged, let it run for a few days after fermentation like the drum digester, it can be attached to the holding tank. The methane will then displace the water that will flow out of the exhaust tube. Once all the water is removed, the valve to the exhaust tube on the holding tank can be closed. The tank is now purged and ready for use.

Parts

Two 275-gallon oil tanks

3/4" copper tubing and fittings

50psig pressure gauge

Length of plastic hose for a site tube

Hose clamps

Valve to fit 4" copper pipe

Length of 4" copper pipe for fill tube

Funnel to aid filling.

Operation of continuous output digester

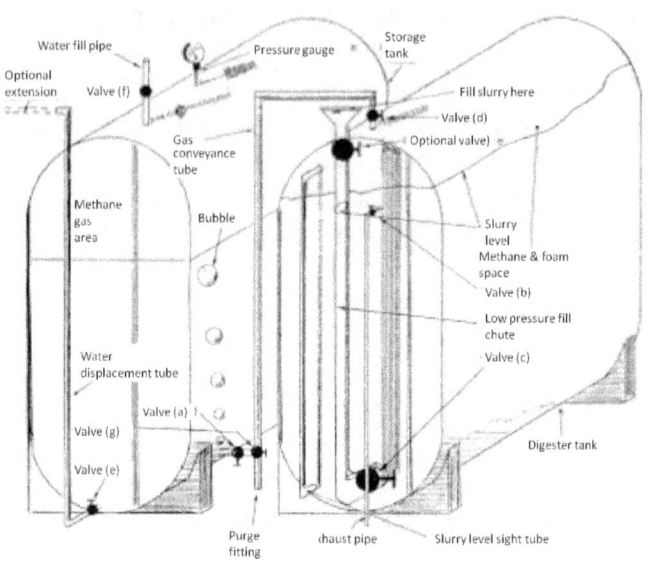

To start the digester close all valves

Open valves (g) (c) and (d)

Fill the low pressure chute with slurry until the sight level tube shows slurry near the top in the tank (one foot space from the top). The slurry will seek its own level therefore there will still be slurry in the fill chute up to the level of the slurry in the digester. The slurry should be allowed to preferment. A cover may be fitted to the top of the chute if desired. Remember to close valve (c) after initial filling and after daily filling.

As gas passes through valves (d) and (g) the system is purged and air is displaced. Fit a hose to the purge fitting and place the other end of the hose in water to check for bubbles of gas. Purge for a few days once the system is fermenting. Make sure no air is present in the line.

Open valve (f) fill the holding tank full with water making sure to expel all the air. Then close valve (f).

Open valves (a) and (e) and let gas enter the holding tank and thus displace the water. It will be forced out of the water displacement tube. An optional extension may be placed on top of this tube, however make sure that the pipe passes above the top of the tank since the water will seek its own level.

A hose may be fitted to the water fill pipe, and the gas consumed by opening valve (f). Close valve (e) once all water is displaced.

Monitor the pressure gauge daily. If the pressure is high in the system gravity will not be enough to push the slurry down the fill chute. A circulation pump would then have to be installed. However the pressure is within the specified range up to 50psig then there should be no problem. To refill daily, close all valves. Open valve (b) and let 1/30th of the slurry come out (this slurry is already digested). Then close valve (b) and replace the 1/30th of the slurry volume through the fill chute.

Further information

The British thermal unit (BTU or Btu) is a traditional unit of energy equal to about 1 055.05585 joules. It is approximately the amount of energy needed to heat 1 pound (0.454 kg) of water 1 °F (0.556 °C). It is used in the power, steam generation, heating and air conditioning industries. In scientific contexts the BTU has largely been replaced by the SI unit of energy, the joule, though it may be used as a measure of agricultural energy production (BTU/kg). It is still used unofficially in metric English-speaking countries (such as Canada), and remains the standard unit of classification for air conditioning units manufactured and sold in many non-English-speaking metric countries.

In the combustion of methane, several steps are involved: Methane is thought to form a formaldehyde (HCHO or H_2CO). The formaldehyde gives a formyl radical (HCO), which then forms carbon monoxide (CO). The process is called oxidative pyrolysis:

$$CH_4 + O_2 \rightarrow CO + H_2 + H_2O$$

Following oxidative pyrolysis, the H2 oxidizes, forming H_2O, releasing heat. This occurs very quickly, usually in significantly less than a millisecond.

$$2 H_2 + O_2 \rightarrow 2 H_2O$$

Finally, the CO oxidizes, forming CO_2 and releasing more heat. This process is generally slower than the other chemical steps, and typically requires a few to several milliseconds to occur.

$$2\ CO + O_2 \rightarrow 2\ CO_2$$

The result of the above is the following total equation:

$$CH_4(g) + 2\ O_2(g) \rightarrow CO_2(g) + 2\ H_2O(l) + 891\ kJ/mol$$
(at standard conditions)

Where bracketed "g" stands for gaseous form and bracketed "l" stands for liquid form.

Biogas websites

We have created a page of biogas links at

https://energybook.co.uk/methane

Please also visit

EnergyBook.co.uk for further publications.

Notes

About the publisher

This book was originally published by OneToRemember. This edition has been published by EnergyBook.

EnergyBook.co.uk

About the author

Richard Jemmett's keen interest in the generation and use of energy has helped to shape his formal career and provide an incentive to write on the subject. He graduated from the University of Leeds in 1980 with a degree in fuel and energy engineering and has held various positions within energy and consultancy companies working in the UK, Europe, the Middle East and Asia. He has been a Board Director of two energy related companies and a past president of the Institution of Gas Engineers and Managers. Richard has written many articles, essays, and conference presentations on utility industry strategy, energy industry development, energy market deregulation, renewable energy, distributed generation and high reliability organisational design.

Disclaimer

The information in this publication has been supplied in all good faith and believed to be correct. However no liability will be accepted for any accident, damage or injury caused as a result, or arising from the use of information from this publication. Information

[i] Methane will not ignite unless mixed with oxygen (air)

[ii] The name natural gas was used to distinguish the gas from town gas that is made from coal or oil. Town Gas was distributed around many cities in the world including the UK, USA and Germany before the commercial extraction of natural gas. It comprises a whole range of gases including methane and carbon monoxide.